Zahra M.M.A. Sadiq

Knowledge Engineering

Artificial Intelligence and Legal Logic

Author Zahra M.M.A. Sadiq

Cover-design Zahra M.M.A. Sadiq

Layout and typesetting Zahra M.M.A. Sadiq

© 2017 Zahra M.M.A. Sadiq

Although the content of this book was developed and compiled with the greatest care possible, errors cannot be completely ruled out. The publishers shall assume no liability due to consequences resulting from incorrect statements.

Anticipating Torricelli's trumpet the Pythagoreans considered the infinite to be limited by the finite • Crito 400 B.C.

Logic is abstract and concrete at the same time because on the one hand it is separable from the reference object and transferrable to another. On the other hand the material and nonmaterial world cannot be imagined without logic • Zahra M.M.A. Sadiq

1. Introduction

By abstracting the structure of different legal systems I compare procedures of case law and codified law. I create an explanatory model of the structure of legal clauses. I develop logical procedures and classify legal language and procedure as a subclass of controlled natural language (CNL).

I describe recurrent neural networks, long-short term memory networks (LSTM) and semi-supervised reinforcement learning (RL) that are applied to design a system using distributed intelligence and knowledge engineering procedures.

Case Law • codified law • abstraction and comparison • creation of a logical legal structure • controlled legal language and procedure • RNN • LSTM • semi-supervised RL • distributed intelligence • knowledge engineering • data engineering

2. The structure of legal clauses

Legal norms that define ethical standards and their application are based on the syntax and meanings of the natural languages. Logic is expressed using specific languages with a particular syntax and semantics and it supports the process of law finding.

I explain the legal reasoning in Common law, Continental European law and Islamic law in a comparative way based on the method of propositional logic or first order logic. I integrate extended models of logic like modal logic and quantum logic to describe specific phenomena that complement the research of the basic logical operations. The significance of logic for legal reasoning is restricted by the specific conditions of natural language and by the qualitative character of the norm. The realm of law is related to the science of philology, linguistics, theology, and the political and social sciences, including philosophy and psychology. Legal reasoning does not exceed the framework of logic which constitutes a procedural mode of thinking.

Although there are obvious similarities between case law, continental European law and legal systems that combine both forms of law, significant differences exist.

Valid legal sources are distinguished from new sources that are developed in order to become enacted law. Concerning the weighting of verdicts and codifications, the legal systems differ. Legal procedure compares real cases and their abstract form. The logical rules are restricted to basic forms, and distinguishing and analogizing are complex scientific procedures that take into account the value of law for society, especially in case an analogy or a teleological reduction is applied.

A logician can neglect concrete cases of the material implication and consider exclusively the abstract conclusion because there are numbers without matter but there is no matter without numbers. The lawyer regards the criteria of truth and falsity on the concrete level of reality and on the abstract level of logic, and he respects the laws of society without breaching any norms of their specific social logic.

Every abstract term designates a number of concrete cases that are imagined or based on real incidents. The development of an abstract model that formulates the conditions of the origin of a legal duty is based on a systematization of single cases and connected with the creation of abstract and transferable terms.

The legal clauses form a net without a hierarchy, and because of their particular qualities, the general legal institution is not the superset of the specific legal institutions. In any legal system based on written law, the basic codification is formed by a constitution that formulates fundamental norms. The constitution is complemented by case law. Two different forms of legal contracts, such as, for example, a streaming service agreement and a contract for the publication of advertisements between an internet academy and the provider of the platform are different by nature. The outcome is that the conditions and consequences of both forms of contracts cannot be transferred from one to the other. Both contracts are disjointed into their consistent components depending on how the contractual obligations and rights are regulated. It is not possible to compare one similar element without respecting the entirety of properties that determine every aspect of the services involved. Because both contracts are conceptually equivalent, a hierarchical order of legal institutions in codifications does not exist.

German law is based on the idea of a hierarchy of legal norms, for example such as that concerning a legal clause and its qualification. This law is constructed by systematic references of legal notions building a hierarchical network of terms and legal clauses. In this system the special term implies the general term. The connection between the legal clauses and the parts of codifications is based on a comparison of the conditions of both norms and the definition of a subset and a superset. From a logical point of view this order of legal clauses is possible but from a legal and semantic perspective, the identity of the sets which are defined by qualitative properties is not provable. This result is also valid for the ramified second order logic, lexicographical orders or other hierarchical logical systems. A ramified second order language is the predicative language L_2K_P that defines relations in levels. Higher order languages can also be ramified. For example, $\forall P_3 \exists Q_1 \forall_X (Q_1X \equiv P_3X)$ means for each level three predicate there is a level one predicate with the same extension. As far as the hierarchy is concerned, relations defined at one level are used in definitions at later levels.[1]

---

[1] Shapiro (2000) 64-65.
[2] Smullyan (1995) 48.
[3] Shapiro (2000) 8.

The legal clauses of a codification are based on the abstraction from a number of concrete cases and objects that are imagined or real. Even if a system is characterized by the assignment of norms to sets forming subsets and supersets, the definition is determined by deductive reasoning, analogizing, and teleological reduction. A large number of possible definitions exists for abstract and general legal clauses, with such a quantity resulting from assimilating cases that differ in quality. Because of this, the conceptual alignment of the legal system in the form of a hierarchical system which consists of generic terms and subtopic terms does not constitute a systematic structure of law.

The structure of legal language is interpreted in a logical way because the form of the legal clause is a material implication with an if-then structure. This form is superficial and it is important for coping with legal language and for analyzing and structuring the internal logic of legal clauses which possess an additive structure. The clause "The user can stream music files" is the basic norm that is extended in the course of history. Later this clause includes additional services, for example the downloading of songs. Commentaries and supercommentaries that depend on each other relate to

6

this rule and refine it. To find this structure it is necessary to analyze the contemporary legal clauses and to reduce them to the basic shortest and most general statement which is extended. This procedure corresponds to the historical development of legal structures and implies the construction of a temporal structure.

3.   Semantics

The truth value of a piece of evidence in the framework of a legal trial determines the logical truth value of the logical statement in the form of a material implication. The legal realm belongs to the social sciences and deals with complex reality. Because of the dual nature of logic, a legal case is regarded from the viewpoint of logic but logic itself is abstract. Logic is abstract and concrete at the same time because on the one hand it is separable from the reference object and transferrable to another and on the other hand the material and nonmaterial cannot be imagined without logic.

In propositional logic, the truth value of a variable is assigned in a direct way. In first order logic, the interpretation of the set of all pure closed formulas of quantification theory designates a function which assigns to each n-ary predicate an n-place relation

(interpretation (term$_1$), ..., interpretation (term$_n$) $\in$ relation r).[2]

Because the legal system is not coherent, the proof of the general validity of a statement is ambiguous. From a mathematical point of view, it is not necessary to prove a theorem by examining every number because this is proved in a general way without examining every figure.

For example a ban with permit reservation designates a legal ban on the investigation of personal data which is only justified in case permission is granted. The relevant administration proves the permission because the burden of proof is reversed. If prohibition has the value

---

[2] Smullyan (1995) 48.

(1), permission has the opposite value (0) and vice versa. It is possible to calculate the truth value of the overall statement.

The ban with permit reservation has the resulting value (0) if the conjunction is

permission (0) ∧ prohibition (1) or

prohibition (0) ∧ permission (0)

The resulting value of the overall statement is (1) if the

prohibition (0) ∧ permission (1)

because permission is the negation of a prohibition and vice versa.

A ban with permit reservation, which is primarily a ban, is the negation of a permit with ban reservation, which is primarily a permit. From the perspective of dichotomizing logic there is no difference concerning the prioritization of the prohibition and the permission, which could be distinguished from an ethical viewpoint. For this gate I do not apply multi-valued forms of logic. It is not reasonable to assign values between 1 and 0 to prohibition or permission.

4.   Legal Logic

Although legal science belongs to the social sciences, the rules and theorems of logic are applicable. Logic consists of a syntax in the form of a language together with a deductive system or semantics interpreting a language or both. [3] Selected axioms which are not provable form the basis of the syntactic system. This system also presupposes a set of rules for inferring new sentences from a finite set of sentences. These axioms are the laws and the inference rules serve to derive the theorems that lawyers call verdicts. The theorem that is proved is the major premise of the syllogism. In some mathematical systems, these rules of inference form a logical self-contained calculus. In the legal realm, the application of the rules of inference including the analogizing and the teleological reduction are based on the lawyer's interpretation and they do not form a calculus.

---

[3] Shapiro (2000) 8.

The development of legal clauses from general basic codification or precedent case is subject to the logical principles of inductive and deductive logic. I use the form of a historical tree to describe the structure and development of the structure of the legal clauses. While the basic legal statement forms the origin, the laws of the second and higher level are the result of deductive and inductive reasoning. All of these legal clauses are phrased in the form of a statement that is specific or general and concrete or abstract. The specific individual legal case forms the basis for inductive reasoning, while the abstract and general clause is the foundation for deductive inference. Deduction and induction form a cycle that starts with solving legal cases by deduction from an abstract and general legal statement. The verdicts that are thereby derived possess a concrete and specific form and are the legal foundation for the inductive creation of new norms.

I presuppose that the rules of the second or higher order are traceable to the basic rule without logical errors, taking into account the systematic context of the framework of the norms and the historical sources. The result is the description of a structurally inconsistent logical system of abstract general legal clauses without a hierarchy of terms.

In the Common Law system, the judges use one ratio decidendi in order to solve the case applying deductive reasoning, and create another ratio decidendi on the basis of inductive inference in the process of formulating a precedent case. A conclusion based on analogizing is an inductive argument.[4] From a logical viewpoint, this process is analogous to the procedure used concerning law that is formulated in an abstract and

The legal system

| Basic law |

Deductive reasoning

| Verdicts |

Inductive reasoning

| Codification |

Deductive reasoning

Repetition of the process for verdicts and codifications

---

4 Essler (1970) 19.

12

general way. Deductive reasoning is applied in the framework of abstract and general legal clauses and inductive reasoning is used for creating abstract clauses.

Analogy of the process of legal reasoning

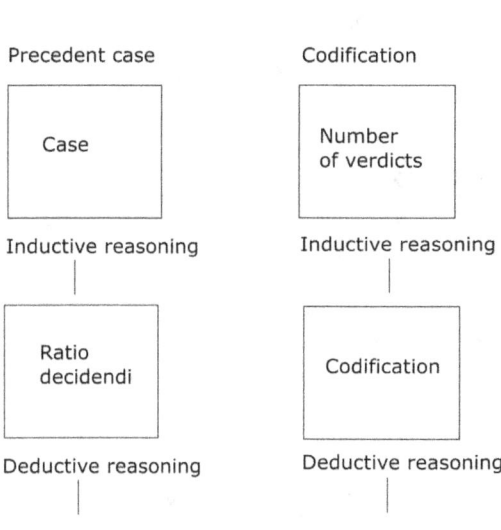

With the exception of basic codification, law develops from verdicts to codifications and from codifications to codifications. The process of creating a codification is not directly reversible because the real legal case is the starting point for the development of law. The deductive application of codifications participates in an indirect way in the solution of cases. Logic is used with different languages that solve different tasks.[5] The classical, multi-valued and modal forms of logic are appropriate languages for the legal realm.

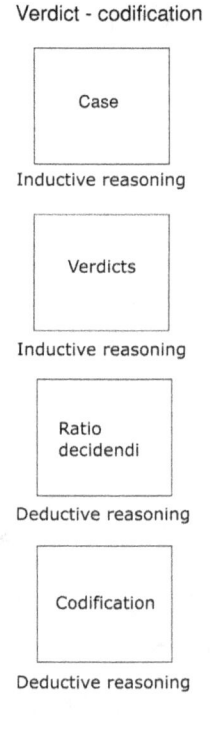

Verdict - codification

Case

Inductive reasoning

Verdicts

Inductive reasoning

Ratio decidendi

Deductive reasoning

Codification

Deductive reasoning

---

[5] Shapiro (2000) 14-15.

a.   Classical concepts of logic

It is possible to apply set theory to define first order logic. Two notions of sets include the iterative set element of set theoretic hierarchy and the logical set which is a collection of iterative sets called a class or a set concept. Logical sets have a Boolean structure. Proper classes which are set classes but not sets are not part of set theoretic hierarchy.[6]

German Law defines general and specific norms and establishes not only a relationship between legal clauses but also between parts of the codification. The general parts consist of universal rules and include rules that apply to the complete codification. The specific parts consist of particular rules and include only rules for the legal principle concerned, for example a particular contract. The single legal clause and its qualification build supersets and subsets.

$$J = M_2 \subseteq M_1 = \forall x \, \{x \in M_2 \Rightarrow x \in M_1\}.$$

---

[6] Shapiro (2000) 18, 177.

The general and the specific set of rules form two sets with a large intersection. The legal principles are different in nature and common traits are connected with the specific legal environment. The analysis of the relations between the legal terms is the task of natural language processing. The legal clause has the form of a material implication. Although codification does not constitute a logical calculus, the logical operations involved are executed regarding the elements of a legal clause. Relations between legal clauses are formulated only to a very limited extent.

A syllogism is used to apply a norm to the legal case in the form of a logical deduction, consisting of a conclusion from the abstract to the concrete statement. In terms of first order logic, the procedure requires the use of a universal and an existential quantifier. The shortened example is taken from section 16 of the Electronic Transactions Act of Singapore concerning the error in electronic communications. The first statement uses the universal quantifier:

$$\forall x\, [Ex \Rightarrow Rx]$$

For every x it is valid that where a natural person (x) makes an input error (E) in an electronic communication

16

exchanged with the automated message system of another party [...], that person [...] has the right to withdraw (R) the portion of the electronic communication in which the input error was made.

One derives the result by applying the universal instantiation.

[Ea] ⇒ [Ra]

If the vendor of an online store (a) makes an error with the pricing of a product (E), he has the right (R) to correct the input error.

The procedure concretizes the properties E and R. The input error (E(x)) is a generic term for the erroneous pricing of the product (E(a)) and the right to correct the input error (R(x)) is a generic term for the right to change the pricing (R(x)). The legal concepts are arranged according to a hierarchical order.

As far as Common Law is concerned, the comparison of the ratio decidendi with the case involves the application of analogizing and distinguishing procedures, thereby concretizing the abstract. From a logical point of view the procedure is analogous to the syllogism. Reasoning from the concrete to the abstract statement reverses the reasoning from the abstract to the concrete statement. The ratio decidendi is formulated by using the inductive method of abstraction from the legal case.

By using existential generalization, I cannot convert the free constant into the bound variable. It is not possible to reverse the procedure. The statement

[Ea] ⇒ [Ra]

is converted into

∃x [Ex ∧ Rx].

This is equivalent to

~∀x [Ex ⇒ ~Rx].

The inductive method concludes general principles and does not presuppose them. Usually the general statement is derived from of a large number of individual cases. The most extreme form of induction is the singular inductive conclusion from a single case that converts the inductive conclusion into a reverse deductive form. [7]

---

[7] Essler (1970) 14-15.

These statements are not called theorems because a mathematical theorem is a universal quantification in a specific field.[8] I use singular inductive reasoning to invert the deductive procedure.

Because of the connection between the general logical conclusion[9] and the singular predictive conclusion[10], I conclude that it is possible to use one singular concrete and particular case to formulate a ratio decidendi. This association is characterized by the abstract general form. Induction leads from the statement to the principle.[11] The existential generalization implies that

$$Pa : \exists x\ (Px).$$

(If a statement is valid for one concrete (a) it is valid for at least one (x)),

is the basis for the derivation of the ratio decidendi. In contrast to this, for example, Article 19 of the German

---

[8] Klüver et al. (2006) 68.
[9] Allschluss.
[10]Singulärer Voraussageschluss.
[11]Essler (1970) 14-17.

Constitution prohibits abstract and general laws based on one specific and individual case. While the statement with the universal quantifier ∀ and the negation of the existential quantifier ¬∃ are not provable, the negation of the universal quantifier ¬∀ and the existential quantifier ∃ are provable.[12]

Analogizing is part of the process of inductive inference. It is a comparison from a subset of properties to another subset of properties. The analogy is represented by the false antecedent and the true consequent of the material implication. Both elements being compared belong to the same superset. Which properties are significant and define the traits that are essential for the nature of the element depends on the context. Some properties which are traits of the original element and its compared element are direct properties and the properties of the context of the elements are properties of secondary degree.

---

[12]Klüver et al. (2006) 32-34.

While the system of Aristotle's deductive logic is coherent, the inductive method is not. First order logic is complete but it is not sufficient for substantial practice.[13] Two of the paradoxes in the framework of induction are the raven paradox, also known as Hempel's paradox and Goodman's new riddle of induction. Both arguments are weak because they require future knowledge.

The reasoner RACE allows the application of generalized quantifiers for collective and distributive plurals. It is implemented as a Prolog program on the base of the model generator Satchmo, which is a theorem prover implemented in Prolog. Collective plurals designate a group of objects and distributive plurals refer to each object of this group. ACE provides five general quantifiers which are divided into three groups. The monotone increasing quantifier designates at least five persons or more than n persons. The monotone decreasing quantifier designates at most five persons or less than n persons. The non-monotone quantifier designates exactly n persons which RACE interprets as

---

[13]Shapiro (2000) 43.

(at least n persons ∧ at most n persons).[14] First order logic is based on constants for properties and individuals and variables for individuals, while second order logic adds the variables for properties. While first order logic is complete, this is debatable for the logic of the second or higher order.

While first order logic is restricted to the quantification of variables for individuals, second order logic also includes the quantification of variables for predicates and function symbols. It is thus possible to express that an individual (x) possesses a property also possessed by individual (y)

$$\forall P \; \exists y \; [Py \Rightarrow Px].$$

Third order logic also includes variables for relations of relations, functions of predicates and functions of functions.[15]

---

[14]Fuchs (2012) 86-87.
[15]Shapiro (2000) 35, 43, 65.

2 2

## b.	Multi-valued concepts of logic

Multi-valued concepts of logic use discrete degrees of truth values between 1 and 0 for statements. The prove procedure is based on the creation of random sets and the research on the conclusion's validity in as broad a number of cases as possible.

Machine learning patterns are based on multi-valued models and its foundation is the algorithmic learning theory based on statistics. [16] Probability calculation is different from fuzzy logic, which deals with possibilities. The truth value of fuzzy logic is multi-valued and the value varies between 0 and 1. According to the rules of fuzzy logic, the set union of A and its complement is not an empty set. Probability is a measure to indicate the certainty of an event that is assumed to occur with a truth value of 0 or 1. For a joint probability distribution over events A and B, the conditional probability of B given A is defined by Bayes' Theorem.

---

[16]Wittek (2014) 11.

Regarding the legal clause, the consequence of the logical equivalence basically occurs with a truth value of 0 or 1. Concerning the so-called "should" or "can" provisions, the rules of probability calculation are applicable and the legal conclusion does not necessarily result from the premise. In contrast to the rules of probability, the principle of maximum probability entropy does not assume that the features are conditionally independent. Entropy is a measure that quantifies the uncertainty of the predicted values of a random variable. Because imprecise values are applied, it is possible to refer to databases which are not representative.[17] Legal analogizing is based on the similarity of expressions of natural language and situations. The Bayesian network requires a conditional probability table (CPT) that is filled in completely and accurately with probability values. The principle of maximum probability entropy is more general than the Bayesian network and makes it possible to fill in probability values that are incomplete. It also allows the use of interval values. In order to process the legal clause in the form of a material implication, both methods

---

[17]Loeckx, https://arxiv.org/abs/1711.01431.

are combined. By formalizing natural language, the principle of maximum entropy contributes to the generalization of the rules.

Probability logic and fuzzy logic are based on extensions of propositional logic. The similarity between two cases, a case and a clause of a codification, or two clauses of codifications that result in the same consequences is the result of an interpretation. This exegesis is characterized by the ambiguity of natural language and requires that the program is able to cope with it. The ambiguities are either based on the meaning of words which are equivocal and the ambiguities of plural forms, or rest upon grammatical references. If the same word is a noun or a verb simultaneously the ambiguity is lexical. The application of multi-valued concepts of logic is appropriate for determining to which extent words belong to the same category or deciding whether two specific terms are synonymous, similar or antonymous. The procedure requires that the value of the relationship is quantifiable.

Reasoning that allows derivations from sets and membership functions with approximate values bridges the gap between natural and formal languages. The multi-valued set whose elements are natural language terms is the sum of the functions that describe the term approximately. The degree of belonging to a set is defined by quantified values. The domain of definition is the set of input values and the output is the image of the function. Sets based on the principle of excluded middle are special cases of fuzzy sets. The probability calculation is used to determine a ranking of legal terms according to their relevance. Multi-valued concepts of logic are also applied to reveal the degree of affiliation of a concept to a cluster of terms. Quantum computing replaces the probability calculation of classical computing with the calculation of amplitudes.

c.   Modal logic

Attempto Controlled English (ACE) is a controlled natural language that provides modal constructs for possibility, necessity and sentence subordination that are represented in the discourse representation structure (DRS). It uses the three modal operators diamond <>, box [] and colon :. By applying the standard translation formulas, modal logic is integrated into formulas of first order-logic.[18]

––––––––––––––––––––––––

[18]Kuhn (2012) 88.

The procedures of deductive and inductive reasoning based on modal logic are combined. The deductive reasoning procedure includes the statement (If A is true, then B is true). The inductive reasoning procedure includes the statement (If A is false, then B is false). Inductive reasoning is denoted by the operator $\Rightarrow^{-1}$, which is a partial order. The rules of transitivity are applicable. In case of a transitive relation, the statements (If A is false, then B is false) and (If A is false, then C is false)

$B \Rightarrow^{-1} A$ and $C \Rightarrow^{-1} A$

are permitted. The proximity induction theorem implies the rule that the risk of induction of the statement (If A is false, then B is false) is not more than the risk of (If A is false, then C is false). The risk of induction is the probability of invalid results. The conditional probability P (~B|A) generalizes the logical statement $A \Rightarrow \sim B$. The risk of (If B is false, then A is false) is the probability of (not B given A).

The risk of $A \Rightarrow^{-1} B$ is $P(\bar{B}|A)$.

Inductive reasoning in the form of the inverse deductive reasoning procedure is applied in the framework of clustering. Incremental clustering is a process of induction from the modal possibility atom to the propositional atom of each data. ζ is a clustering algorithm. ϕ is a modal language. Each element of ϕ_ζ is a propositional atom. φ is a modal formula of ϕ_ζ. According to the theory of modal logic and Axiom B from the "system B"[19], (it is true that if it is necessary that φ is possible then φ is true)

$$\Box\Diamond\varphi \longrightarrow \varphi.$$

(If d is true, it is true that if P is true, then it is necessary that P is possible).

$$d \Vdash P \longrightarrow \Box\Diamond P$$

There exists a deduction relationship by modus ponens from the statement (If d is true then P is true to if d is true, then it is true that it is necessary that P is possible)

$$d \Vdash P \text{ to } d \Vdash \Box\Diamond P.$$

---

[19]Hughes et al. (1996) 63.

There exists an induction process from the statement (If d is true, then it is necessary that P is possible) to (If d is true then P is true)

$d \Vdash \Box \Diamond P$ to $d \Vdash P$.

Because Axiom T implies that

$\Box \varphi \longrightarrow \varphi$.

The cluster P is a subset of the cluster $\Diamond \varphi$. The sentence $\Diamond P$ is interpreted as "close to cluster P". $\Box \Diamond P$ is far away from $\sim \Diamond P$. $\Box \Diamond P$ is near to P but it is less near to $\Diamond P$. D is a finite data set. The similarity is a binary relation S on the data set D

$S \subseteq D \times D$.

The data d_1 and d_2 are elements of S and they are similar according to the Euclidean distance. The similarity function that constructs the Euclidean distance is reflexive and symmetric. Each element of ζ (D, S) is called a cluster. ϕ is the modal language and the cluster $P \in \phi$. $\Diamond P$ is a fuzzy cluster. ζ is the clustering algorithm. [20] This procedure is connected with modal language and it is transferred to other languages of logic.

d.   Quantum logic

Quantum computing is based on various models which include the topological quantum computing, the one-way quantum computing and the adiabatic quantum computing that uses Hamiltonians to manipulate states. In quantum mechanics the Hamiltonian is an operator that corresponds to the energy of the system and it is a function over a space of physical states to another space of physical states. By defining operations on single or multiple qubits, quantum circuits are constructed.[21] This method converts the state of superposition of quantum bits into a classical state by measuring. During the superposition, the state of the qubit is expected to be 1 and 0 at the same time, and after the measuring it is 1 or 0. The result of the measuring depends on the amplitudes α and β.

---

[20]Zongle et al. (2009) 1132-1143.
[21]Wittek (2014) 41, 44.

If a qubit in the state $\alpha \cdot | 0\rangle + \beta \cdot | 1\rangle$ is measured, it is in the state $| 0\rangle$ with a probability of $|\alpha|^2$ and in the state $| 1\rangle$ with a probability of $|\beta|^2 = 1 - |\alpha|^2$. The state of the quantum bit is described as a vector with a length of 1 in a two-dimensional complex vector space. If the state vector is multiplied by a unitary matrix it is transformed into a new probability distribution vector whose elements sum to 1. This transformation is based on the inverse Hadamard matrix. States of a quantum register with n bits are vectors in a 2^n-dimensional complex vector space. The Tensor product is the result of the calculations of the single bits $V_1 \otimes V_2$.[22]

Quantum logic uses specific algorithms and quantum gates, and classical circuits are transformed into quantum circuits. For example the CNOT gate is the unitary variant of the exclusive OR gate. If the first control qubit is 1, a NOT operation is applied on the second target qubit and if the first bit is 0, the second bit is unchanged

CNOT $|x, y\rangle \mapsto |x, y \oplus x\rangle$.

---

[22]Homeister (2013) 20-30, 52.

It is possible to replace each gate of a classical logical operation with two input bits and one output bit by Toffoli gates, which are also named CCNOT. [23] If the first two control qubits are set to 1, the third target qubit is changed, otherwise all bits stay the same.

CNOT $|x, y, z\rangle \mapsto |x, y, z \oplus x, y\rangle$.

e.   Temporal and spatial forms of logic

The interrelated elements of time and space are integrated into the reasoning procedure.[24] This temporal and spatial relation of logic is part of the learning algorithm that enables the program to make choices and to select. The temporal dimension of the material implication consists of a time interval (i) and a moment (t). The state (s) exists in the interval (i) or at the moment (t) or it is exclusive. The temporal and spatial relation implies that (if the state (s) exists in an interval (i), the

---

[23]Homeister (2013) 88-93.
[24]Shi (2011) 163.

state exists at the beginning of the moment (t) and at its end). The state with this property is called the position state.

$\forall_i$　(Holds-on(s,i))　$\longrightarrow$　Holds-at(s,inf(i))∧Holds-at(s,sup(i))

Holds-on(s,i) designates that state (s) exists in interval (i). Holds-at(s,t) designates that state (s) exists at moment (t). Div(t,i) designates that moment (t) is in interval (i). sup(i) designates the ending moment of interval (i).

The motion state has the property of designating that (if the state (s) exists at moment (t), there must be an interval containing this moment which (s) exists on).

$\forall_t$ (Holds-at(s,t)) $\longrightarrow \exists_i$ (Div(t,i)∧Holds-on(s,i))

It is possible to reason with the spatial and temporal procedure.

DC(a,b) exists at moment (t_1).

PO(a,b) exists at moment (t_2) and $(t_1 < t_2)$.

In this case there must be a moment (t) in (t_1, t_2) at which EC(a,b) exists.

Holds-at(DC(a,b),t_1)∧Holds-at(PO(a,b),t_2)∧t_1<$t_2$$\longrightarrow$
t(div(t,(t_1,t_2)∧Holds-at(EC(a,b))t)

If two regions are not in touch with each other at moment (t_1) and two regions partly cover each other at moment (t_2), and (t_1) takes place before (t_2), then moment (t) exists in the moments (t_1, t_2) at which two regions are in external touch with each other. $DC(x,y)$ designates that two regions are not in touch with each other. $EC(x,y)$ designates that two regions are in external touch with each other. P is the transitional state predicate. $P(x,y)$ designates partly belongs to. O designates cover each other. The relation is $C(x,y)$. Because P is a fuzzy expression, this method is combined with multi-valued forms of logic or modal logic. The moment (t) and the temporal interval (i) is defined with a precise or a fuzzy value.

f. Unidentified logical variables

Since the universal quantifier $\forall$ symbolizes a chain of conjunctions the existential quantifier $\exists$ symbolizes a chain of disjunctions. The negation of this quantifier follows the laws of De Morgan, which are symmetric rules of equivalence. The negation of the universal quantifier implies

$\neg \forall x\, [Ax]$ corresponds to $\exists x\, \neg[\,Ax]$.

The negation of the existential quantifier implies

$\neg \exists x\, [Ax]$ corresponds to $\forall x\, \neg[\,Ax]$.

3 4

If the number of variables that the logical statements imply is unknown, the truth value of the complete statement is undetermined. This is valid for the number of variables of the premise or the consequence of a legal material implication if it implies a general clause or an undetermined legal term or uses an analogy or a teleological reduction. The paradox of the ship of Theseus and the Sorites paradox are examples of ambiguities. If the premise is true, the consequence is also true or vice versa because in law basically logical equivalencies are applied. The enthymeme consists of a simulated assumption that no premise was neglected which is necessary in order to prevent the argument from being invalid.

g.   Logical continuation of errors

As far as the derivation of the ratio decidendi or the development of the law is concerned, an error causes consequent errors if there is a chain of derivative reasoning. In coherent legal systems, this continuity lasts for several centuries. If a court or the Supreme Court controls a legal clause, it examines the accordance of the law with the superior law constituting the legal foundation. If it is not possible to relate the legal clause to the basic norm in a direct way, the chain of derived clauses is integrated into the reasoning procedure.

5.   Distributed artificial intelligence

Procedures that integrate the computer into the process of legal reasoning apply distributed artificial intelligence, natural language processing, and neural networks. The cooperation of the computer with the legal expert is semi-automatic. The legal system refines its own terms and a system of terms and rules and adapts itself to the changing environment. Distributed Intelligence systems improve legal reasoning in a complex reality. The application of distributed intelligence systems integrates the human expert into the training of the data and the evaluation of the testing procedure.

In the course of history, the processing of legal clauses has had an additive structure. Starting with a general rule, special cases and features are supplemented. This development has a scientific objective when the items in the literature refer to one another in the form of commentaries. Based on the research into the development of legal clauses, one applies logical reasoning to transform a legal clause in one time period into a second legal clause in the subsequent period. If the distributed intelligence system is used to solve legal cases, verdicts or legal clauses of codifications that have been overruled are not included.

If the legislator wants to find a new norm in a new context he applies parsers that are based on look-ahead features or he uses RNNs to find out how the beginning of a text is continued. This procedure requires a form of controlled natural language (CNL) that is specifically defined. [25] I define formal reasoning as a part of the controlled legal language processing (CLLP), which is a combination of CNL and controlled legal language (CLL). The software learns to develop clouds of terms and to apply rules of inference developing new solutions.

The development of law is a complex process based on the decisions of parliament, which enacts codifications, and the population, which complies with the rules. Legal distributed intelligence systems learn to cope with the complexity of the legal sciences. This software does not limit the multilayered and often ambiguous

---

[25]Kuhn (2012) 97-99.

considerations a judgment is based on. It regards the structure of the law from a macro-level and a micro-level perspective and considers the compatibility with the sociological circumstances of law. Legal distributed intelligence systems contribute to the processes of legal application and law-giving. This procedure is meaningful for the development of legal clauses referring to the technical realm because it facilitates the integration of specialist knowledge into the process of lawgiving. Complex neural networks are combined with multi-valued methods. If the input concerns specific solutions, the user invests time in the evaluation of the answers of the program and considers the legal clauses, verdicts and commentaries.

a.   Legal Expert System

Legal expert systems (LESs) present knowledge in the knowledge base. Logical rules generated from the knowledge of experts, knowledge engineers and databases are integrated into the knowledge base. This knowledge is translated into formalized languages, for example propositional logic, first order logic, fuzzy logic, probabilistic logic and decision trees. The structured or unstructured information that is stored in the knowledge base is used by the computer and the inference engine decides how the logical rules are applied. The inference methods are separated from the knowledge base in order to be transferable to a different knowledge base. Artificial intelligence procedures need an extensive quantity of information in the knowledge base.

b.   Knowledge Engineering

Knowledge engineering integrates domain experts into the process of creating software. Researchers of specific realms other than information technology participate in the process of creating distributed intelligence systems and cooperate with the coder. The knowledge engineer processes scientific thinking and combines it with the formal language of computer science.

Knowledge engineering presupposes the thematic evaluation of the subject. I shape a conception of legal science that depends on the analyzed structure and adapts it to the method of information technology. Ontology design and data design are conditions for structuring the data according to systems. They contribute to the application of the ontology to the legal data. The model is the analysis of historical law according to the logical rules and the temporal structure of legal clauses. This system is abstracted and transferred to contemporary law in order to find an inductive procedure to determine structures of the future development of law.

c.   Ontology engineering

The ontology expert creates methods of building ontologies and transforms the particular knowledge into rules and ontologies. Knowledge acquisition and documentation structuring (KADS) is a method that is applied to describe the development of organizations.

Knowledge engineers build ontologies that imply sets of concepts in a domain and the relationships between those concepts. The conceptualization is transformed into machine-interpretable ontologies which provide structured data based on RDF, RDFS or OWL. Ontological axioms are transformed into statements of description logics (DL) which are a family of formal knowledge representation languages. DL possesses a temporal dimension and the logical form is multi-valued. Axioms of description logic are translated into a serialization of RDF in the form of RDF/XML or Turtle or

SWRL rules. Concept definitions are mapped to resource segments of the RDF and linked to related resources of knowledge bases, ontologies and LOD data sets.[26]

d.   Design of the agents

The ontology helps to design the distributed artificial intelligence system and multi-agent-system (MAS). Distributed intelligence divides data and knowledge logically and physically. The cost of communication between the different instances is less than the cost of the problem solution. Legal information mining and intelligent agency are combined based on information theory.

The cognitive sciences contribute to the development of information technology and cognitive computing. Symbolic intelligence and neural computing are combined. Probabilistic algorithms and computational information theory that measures the amount of information required to process CNL computationally improve processing time. The software agent is part of a

---

[26]Afriyanti, https://arxiv.org/abs/1707.02511.

program that shares and receives information and the natural language is translated into symbol description. The agent processes information, performs its specific function and contributes to the processing of the entire tasks. The knowledge of the provider of the information is useful in a particular period of time. Reactive agents focus on meeting the requirements in a dynamic environment. In the context of NLP, I prefer the deliberative agent that is a knowledge based system and which requires the translation of descriptions of the real world into the programming language. Deliberative agents perform better in proving their plans mathematically and they possess a world model in the form of complex symbol reasoning. The concept of the deliberative agent is abstract and it refers to symbol information processing. The knowledge base of the deliberative agent includes both general and real knowledge.

The most commonly used architecture is the belief–desire–intention software model (BDI) that is used to program intelligent agents. The input of the deliberate agent algorithm is the static description of the current world environment, the knowledge base and the plan. The method implies the update of the mental state of the agent according to the update of the perceived world model. The agent plans the future and makes a decision. It evaluates the state, compares it with the knowledge base and chooses an action. After the second update of the world model the agent takes action.

The BDI algorithm starts with initializing the state. Repetition involves the options, the update of the intentions, the update of external events, and dropping unsuccessful and impossible attitudes. The algorithms ends with the command "end repeat". The agent model communicates with the environment and analyses it. In the legal realm, the knowledge base database is partially identical with the environment. The environment includes the existing real cases and the science of law.

A similarity function compares two existing input data sets. In the event of solving a legal case the deliberative agent processes information by deduction and outputs a new rule which is the result of comparing the existing legal clause with the case. The optimization of an existing model of the world predicts unknown or future cases or word formations based on the probabilities that are determined by the training data. The agent shapes its successful handling of its environment by trial and error. In the event of creating a new legal clause, the deliberative agent processes information by induction and outputs a new rule or a set of rules as the result of comparing it with existing legal clauses. The legal language supports the communication and the exchange of information between different cases and rules. The legal language is a subclass of NL that I call natural legal language (NLL).

Agents are as independent as possible and they are not coordinated by a central instance but by their inner structure. Each agent has a linguistic instance that is semantic and syntactic, and a mathematical-logical instance. Each of these instances possesses an intrapersonal instance that regulates processing, and an extra-personal instance that is competent for the exchange. The agent possesses intrapersonal, extra-personal, and linguistic and mathematical-logical intelligence. Intrapersonal intelligence concerns goal-directed activity and decision making. The agent sets systems and adapts them to the changing environment. Extra-personal intelligence refers to the interaction between negotiation and delegation. Technical intelligence is linguistic and mathematical-logical.

4 4

Agents communicate by using network protocols like NFS and HTTP. The information exchanged includes the format of the information and the message type related to grammar rules. The content language possesses propositions, objects and actions. Propositions indicate whether a sentence is true or false.

An item in a data set has different meanings that refer to the perspective of the description. Expectations in the form of probability and possibility depend on the context. The information content of the data is ambiguous and certain meanings are more probable if particular contexts are noticeable. The application of the interleaving of mathematical functions in the framework of neural nets or RL contributes to the solution.

e.   Partial Global Planning

Different agents perform their tasks simultaneously without interacting and exchange information with other agents. Does the society of artificial agents need external rules which coordinate the cooperation of the agents? The agent society learns, teaches and improves. Past experiences shape the decision-making processes and they are helpful if the agent receives a reward or a cost when appropriate. This procedure can also impede good decisions if costs and rewards are disproportionate. It regulates the reward and cost system of each agent and of its community. If one agent studies, learns and applies its knowledge, it acts in the current state. It evaluates the past and plans the future for its present on the strength of the reward or cost available in the problem solution process. Agents optimize their strategy by applying the Partial Global Planning framework.

* Each agent creates the partial planning
* The agents exchange rules
* The agents restart the partial planning
* The agents revise and optimize their strategy

Distributed partially observable Markov decision problems (POMDPs) are appropriate for quantitative performance analysis in the case of uncertainty. The method is agent-oriented and object-oriented. The agents communicate by using the Knowledge Query and Manipulation Language (KQML), which is a language and protocol for communication among software agents and knowledge-based systems. [27]

---

[27]Shi (2011) 499-552.

6.   Methods of NLP

"As a mathematical model, there is always a gap between the language of a logic and its natural language counterpart (...). For example it is sometimes difficult to find a formula that is a suitable counterpart of a particular sentence of natural language, and moreover there is no acclaimed criterion for what counts as a good, or even acceptable 'translation'."[28] In the realm of mathematics, the designations of sets and properties of numbers are quantitatively identifiable while the natural language is ambiguous.

The unification of legal speech is used in order to avoid the ambiguous formulations of natural speech. A noun which is interpreted as a verb causes lexical ambiguity. Languages like the "Controlled Legal German", which is based on deontic logic[29] or "Attempto Controlled English", standardize and specify natural language. The

---

[28]Shapiro (2000) 3.
[29]Höfler et al., https://files.ifi.uzh.ch/hoefler/
hoeflerbuenzli10ifitechreport201011.pdf.

formalization of legal speech and specifically of ambiguous formulations constitutes a precondition for future automatic processing. Specific legal material implications are often based on narrations of parties in a legal process. The abstract formulation of the legal statement uses complex speech, which is interpreted by the lawyer who concretizes the general form.

Artificial intelligence procedures apply neural networks. Quantum neural networks replace weights with entanglement.[30] Machine learning techniques are based on the development of successful proofs and the heuristic evaluation of errors and advances. Because of the default logic it is possible that properties of objects are valid until divergent rules apply. The usefulness of these methods is limited by semantic and practical problems.

Definitions of complexity are based on entropy and the information contained in a chain of strings. Entropy designates the probability of the occurrence of a specific

---

[30]Wittek (2014) 69.

48

order, for example, of a string. The information content of a system of strings is dependent on the comprehension of the receiver.[31]

In order to develop new laws, it is necessary to discover a legal gap. Because there is no superset, the existing norms do not form subsets or complement sets. In the process of comparing real cases with existing norms, the program copes with CLL, which is a particular form of CNL. The input of a chosen selection of words is compared with a cloud of words using databases grounded in online thesauri and vector-based dictionaries, including the web based dictionaries. This strategy contributes to the discovery of related terms and it finds the term with the special meaning resembling the term which is sought after. For example the term phishing is related to spying out data although both actions do not breach the same law. If the number and relevance of the unsolved cases is significant, the legal gap is undesirable and the program formulates a new norm. This procedure is based on the abstraction from

---

[31]Klüver et al. (2006) 280-281, 300-306.

existing real cases. As far as criminal law is concerned, and outside the context of other laws, the principle of definiteness demands that the program finds a rule which is distinct from the existing rules.

a. Neural networks

The recurrent neural network is a form of input-output mapping network which is preferred to create a dynamic temporal behavior for sequential data. The plot of the narration and the additive structure of the legal clauses imply a temporal structure.

The definition of vectors and weights for the development of a neural net based on recurrent neural networks (RNNs) requires time management. The output consists in terms and phrases with a conceptual similarity. Resting on the analysis of the past, the legal clause is gradually complemented. If the training data consist of legal terms, sentences and cases, one finds general terms for classified groups of words which often appear together in a legal text. The relationship between the output and the input is determined statistically.

50

Neural machine translation (NMT) procedure does not divide the input sentence into phrases and words using statistical methods but refers to the entire sentence. The system is based on two recurrent neural networks (RNNs). They consist of the encoder network for the input text and the decoder network of the output text. In contrast to conventional RNNs the long short-term memory network (LSTM) enables neurons to consider effects that are temporarily delayed and excludes the vanishing gradient problem. The input gate, the forget gate and the output gate of the LSTM are perceptrons. This procedure requires great computing capacity. The processing of back propagation changes the weights of the cell state itself. Applying NumPy, the programming procedure consists of four steps. The first is to build an RNN class, the second is to build an LSTM cell class, the third is to program the data loading functions and the fourth is to train the model. The first part of the procedure applies text recognition techniques with an encoder RNN input and output. The second part is the textual production with an encoder RNN input and output. It is important that the RNN has the ability to generalize.

Later laws cancel former ones. This principle of "lex posterior derogate legi priori" is part of the dynamic process of legal development. The Hopfield network, a recurrent neural network, is an appropriate method for programming this form of overruling because the inactive nodes are dropped from the set of active nodes. The Hopfield network is a simple group of perceptrons that solve the XOR problem.

Deep learning procedures are artificial neural networks which apply more than one hidden layer that automatically form a hierarchy. From a layer with a low level of abstraction other layers with a higher level of abstraction learn. [32] Deep learning combines unsupervised and supervised learning methods.

The method of using word vectors or thought vectors[33] is based on integrating numbers that represent a word. Thought vectors designate the meaning of words that belong to natural language and legal language. By using

---

[32]Hinton, Geoffrey et al., https://www.cs.toronto.edu/~hinton/absps/DNN-2012-proof.pdf.
[33]Dai et al., https://static.googleusercontent.com/media/research.google.com/de//pubs/archive/44267.pdf.

an encoder RNN, a sequence of words is converted into a "thought vector". Each word is converted into a vector of features designating the meaning of the word and its context. The word is embedded in a vector space using a neural network like the word2vec model that learns to generate its context through repeated guesses and that is built in TensorFlow. [34] The decoder of an RNN outputs probabilities for possible words first. After the first word is found, the RNN specifies probabilities of other words.

b.　Supervised Learning of CLLP

By analyzing legal texts, knowledge about the language involved, the words and grammar used in verdicts or legal documents, is improved. In the process of training the program, the legal expert chooses the most appropriate terms because the subtleties of natural language are important and because the user of the application expects a particular output.

The forms of linguistic analysis are semantic, pragmatic, syntactic and lexical. The syntactical analysis examines the structure of the sentence. Syntactical languages are

---

[34]https://www.tensorflow.org/tutorials/word2vec.

formal. The computer is able to process the syntax of the natural language if it is formulated in a standardized form which uses Controlled Legal Languages (CLLs). The semantic and pragmatic analysis considers the meaning and the contextual application of the language. The relationship between the signs, the character strings and their meaning is semantic. The connection between the signs, the author, the sender and receiver comprises the pragmatic aspect.[35] As far as the semantic processing of the language is concerned, the program searches for properties of the elements of the sets that are defined to be key terms. The hierarchical system defines sets of words that comply with an umbrella term. If the degree to which the term belongs to one set or another is quantifiable, multi-valued forms of logic are applicable. Lexical analysis is the process of developing a lexicon or multiple lexicons and it is combined with the word2vec method. [36] This word embedding method is used for semantic similarity calculation.

---

[35]Schweighofer (1999) 11, 105.
[36]Sugathadasa, et al., https://arxiv.org/abs/1706.01967.

In order to analyze texts written in natural language, it is possible to separate nouns and verbs from conjunctions and to describe word order and spelling changes. Stop words, for example, "the" and "that", which appear frequently, do not possess a significant meaning. Synonyms and homonyms are processed according to their meaning. Nouns, adjectives, verbs and conjunctions are divided into the category important or unimportant by applying the classical Bayes network. Morphological analysis involves each inflected or derived word being reduced to its stem by lemmatization. Stemming removes the prefixes and the suffixes and takes into account words that have similar spelling.

A grammatical rule is represented by significant terms and is formulated in Controlled English.[37] This method depends on the language that is used. The training of the descriptive grammar of a natural language is based on the analysis of the position and scope of anaphoric references and on determining it statistically. In the first phase the biphasic hybrid system applies supervised

---

[37]Dantuluri et al. (2012) 57.

learning and in phase two the program transfers the knowledge in order to adapt to new environments using techniques of reinforcement learning (RL).

c.   Unsupervised learning of CLLP

If an integrated database is used for the division of words into classes, a classification algorithm is applied. Nouns, adjectives and verbs are classified according to their content-related meaning by applying unsupervised learning procedures and a clustering method. Clusters are groups formed of data instances with similar distances. The more limited the distance in a multidimensional feature space is, the more similar are two examples. A certain abstract term might refer to a variety of sub-concepts with very different and even contradictory meanings. The classification of documents is implemented by using hierarchical clustering. [38] Quantum divisive clustering is the quantum variant of divisive clustering and it is a form of hierarchical

---

[38]Sahoo, http://www.cs.cmu.edu/~ callan/Papers/cikm06-nsahoo.pdf.

clustering.[39] The comparison of notions depends on their conceptual context. The distance is determined by applying the classical K-means clustering algorithm, or hierarchical clustering methods. These forms of unsupervised learning detect simple geometric structures. If the input and output vectors are quantum states, the speedup of the quantum K-means algorithm which uses Grover's search algorithm over the classical variant is feasible.[40]

More complex geometrical structures of the data sets are detected by density-based clustering that is similar to manifold learning methods. The centers of the cluster are optimized by using the iterative expectation–maximization (EM) algorithm.

d.   Reinforcement learning of CLLP

Q-learning is a reinforcement learning technique that is based on a trial-and-error procedure and that finds an optimal strategy by learning an action-value function. The

---

[39]Wittek (2014) 60-61, 104-105.
[40]Wittek (2014) 123.

Q-learning algorithm is applied to find the optimal policy for Markov decision processes (MDPs). The MPD is a mathematical framework that is used to model decision making processes. Reinforcement learning starts with the demonstration by natural language recordings or by programming. The expected value of all possible subsequent stages of an action is calculated by applying the probability calculation in order to determine the Q-function. The embedding vectors for the text are divided into relevant and irrelevant groups which results in a higher Q-function value.

Natural language text strings describe actions whose space is discrete and potentially unbounded because of the complexity of the natural language. The network trains state texts to describe scenes and action texts to describe potential actions and measures their relevance to the current context. Applying deep neural networks (DNNs), the action (a_t) and the state (s_t^i) are mapped to their embedding vectors, and by using an interaction function, the Q-function is approximated. One selects the optimal action (a_t) for a particular state (s_t) by finding the maximal value of the Q-function and by applying the arg max function.[41]

RL is applied in a semi-supervised form. One trains an RL policy in settings where a reward function is available. An inverse reinforcement learning algorithm is applied and the agent simultaneously learns a reward and a more general policy for unlabeled data. The agent develops a policy in the labeled MDPs and the supervisor trains a reward function with supervised learning. This reward function is applied for learning in the unlabeled data. Unlike the policy, the reward function is decoupled from the rest of the MDP and therefore generalizes more readily. The algorithm alternates between inferring the reward function and updating the policy.[42] The standard method of reinforcement learning consists in learning a policy in the labeled MDPs and applying it directly to new MDPs from the same distribution. The agent is provided with reward functions in a limited set of situations that is trained by a human

---

[41]He et al., https://arxiv.org/abs/1511.04636.
[42]Finn et al., https://arxiv.org/abs/1612.00429.

supervisor. The reward function of semi-supervised reinforcement learning is evaluated in a small set of labeled MDPs. The resulting policy and the reward function must be successful on a larger set of unlabeled MDPs. Semi-supervised deep RL uses experience from the sets with only partial access to the reward function. The intelligent agent receives a partial supervisory feedback that is based on an algorithm. This algorithm transforms the conceptual knowledge that is generalized and transferred to new unseen conditions. The procedure regularizes side information that comprises the structure of the data. The technique resembles the inverse reinforcement learning algorithm (IRLA) and applies the reward function to the unlabeled settings by learning from the successful trials in the labeled settings.

e.   Quantum learning

Within the framework of classical learning procedures, semi-supervised learning and active learning, which is a variant of semi-supervised learning, are inductive. The learning algorithm of active learning infers a function from labeled or unlabeled data points to unseen data points from an information source. Instance-based learning, for example in the form of K-means clustering, is also inductive. It infers from the particular to the general and finds general rules to label future data.

The selection of the optimal state in quantum process tomography uses transductive learning, infers a function and applies it from the particular to the particular. The quantum learning process in the form of transduction avoids inductive learning. Its incoherent strategy, similar to the inductive learning technique, is based on detaching the unitary from the examples and storing it in classical memory. This incoherent process induces a function that is applied to any example but because the approximation function is restricted to unitaries, the application of the method is limited.[43]

7.   Complexity of NLP

A strategy that decreases complex structural dependencies is used to reduce the complexity of natural language processing and to support tractability. It facilitates the structured prediction of words and

---

[43]Wittek (2014) 15-18, 135-139.

improves generalization. In order to advance the predictability of words, one decreases complex structural dependencies to reduce the error rate, noise and over fitting. This is achieved by using a simple structural complexity regularization solution based on structure regularization decoding (SR Decoding). The complex structure is regularized by the simple structure to improve the generalization ability of the model. This method is applied to multiple sequence labeling tasks using a linear-chain model that can consist of BLSTMs. A hierarchical model like the perceptron model is used for the parsing task.

The method applies structure regularization decoding algorithms for several important natural language processing tasks that include sequence labeling tasks and parsing tasks. Sequence labeling tasks include chunking and name entity recognition and the parsing tasks include the joint empty category detection, phrase-structure trees and dependency trees. The sequence labeling task consists of finding a label sequence with the maximum probability. The procedure aims at estimating the joint probability of the labels and the sequence of words that is the condition. The number of parameters that is estimated for modeling the joint probability is extremely large and this makes the problem intractable. Markov assumptions reduce the parameters and make the procedure tractable. Dependency parsing aims at predicting not only the dependencies among normal words but also the dependencies between a normal word and an empty element. Structural complexity regularization decomposes the dependency scope of the training samples into smaller localized dependency scopes that form small samples for the learning algorithms. For achieving the structured prediction, structural complexity regularization is as important as weight regularization for reducing the generalization risk.[44]

---

[44]Sun et al., https://arxiv.org/abs/1711.10331.

8.   Summary

I structure a net of legal clauses that does not form a hierarchy. Because of the particular qualities of the norms, I define that the general legal term is not the superset of the special expression. I explain the development of legal clauses from codification or precedent case that is subject to the logical principles of inductive and deductive logic and that form a cycle. I combine the procedures of deductive and inductive reasoning based on modal logic and substitute the modal logic by multi-valued forms of logic.

I apply ontology aids to design a distributed artificial intelligence system and multi-agent-system (MAS). Distributed intelligence divides data and knowledge logically and physically. I regard the ontology design and data design as conditions for structuring the data according to systems. They contribute to the application of the ontology to the legal data. By regarding legal language as a special form of CNL I avoid the combined complexity of procedures.

Numbers are the essence of all items • Pythagoras

Index

Bibliography

Afriyanti, Lis et al., Feature Model-to-Ontology for SPL Application Realisation, https://arxiv.org/abs/1707.02511.

Dai, Andrew M. et al., Semi-supervised Sequence Learning, https://static.googleusercontent.com/media/research.google.com/de//pubs/archive/44267.pdf.

Dantuluri, Pradeep et al., Engineering a Controlled Natural Language into Semantic MediaWiki, in: Controlled Natural Language, ed. by Rosner, Michael et al., Berlin et al. 2012, 53-72.

Essler, Wilhelm K., Induktive Logik. Grundlagen und Voraussetzungen, Freiburg, München 1970.

Finn, Chelsea et al., Generalizing Skills with Semi-Supervised Reinforcement Learning, https://arxiv.org/abs/1612.00429.

Fuchs, Norbert E., First-Order Reasoning for Attempto Controlled English, in: Controlled Natural Language, ed. by Rosner, Michael et al., Berlin et al. 2012, 73-94.

Gardner, Howard, Frames of Mind. The Theory of Multiple Intelligences, New York 2011.

He, Ji et al., Deep Reinforcement Learning with a Natural Language Action Space, https://arxiv.org/abs/1511.04636.

Hinton, Geoffrey et al., Deep Neural Networks for Acoustic Modeling in Speech Recognition. The Shared Views of Four Research Groups, IEEE Signal Processing Magazine, 29 (2012) 82-97, https://www.cs.toronto.edu/~hinton/absps/DNN-2012-proof.pdf.

Höfler, Stefan et al., Controlled Legal German 1.0. Einführung und Spezifikation, https://files.ifi.uzh.ch/hoefler/hoeflerbuenzli10ifitechreport201011.pdf.

Homeister, Matthias, Quantum Computing verstehen. Grundlagen, Anwendungen, Perspektiven, third ed., Wiesbaden 2013.

Hughes, George Edward. et al., A new introduction to modal logic, London 1996.

Jain, Ashesh et al., Structural-RNN. Deep Learning on Spatio-Temporal Graphs, https://arxiv.org/abs/1511.05298.

Klüver, Jürgen et al., Mathematisch-logische Grundlagen der Informatik, Bochum 2006.

Kuhn, Tobias, Codeco. A Practical Notation for Controlled English Grammars in Predictive Editors, in: Controlled Natural Language, ed. by Rosner, Michael et al., Berlin et al. 2012, 95-114.

Loeckx, Johan, The Case for Meta-Cognitive Machine Learning. On Model Entropy and Concept Formation in Deep Learning, https://arxiv.org/abs/1711.01431.

Sahoo, Nachiketa, Incremental Hierarchical Clustering of Text Documents, http://www.cs.cmu.edu/ ~callan/Papers/ cikm06-nsahoo.pdf.

Schulz, Klaus U., Why combined Decision Problems are often intractable, in: Frontiers of Combining Systems, ed. by Kirchner, Helene et al., Berlin et al. 2000, 217-244.

Schweighofer, Erich, Legal Knowledge Representation. Automatic Text Analysis in Public International and European Law, The Hague et al. 1999.

Shapiro, Steward, Foundations without Foundationalism. A Case for Second-order Logic, New York 2000.

Shi, Zhongzhi, Advanced Artificial Intelligence, Singapore 2011.

Smullyan, Raymond M., First-Order Logic, New York 1995.

Sugathadasa, Keet et al., Synergistic Union of Word2Vec and Lexicon for Domain Specific Semantic Similarity, https://arxiv.org/abs/1706.01967.

Sun, Xu et al., Complex Structure Leads to Overfitting. A Structure Regularization Decoding Method for Natural Language Processing, https://arxiv.org/abs/1711.10331.

Tensorflow, https://www.tensorflow.org/tutorials/word2vec.

Wittek, Peter, Quantum Machine Learning. What Quantum Computing means to Data Mining, San Diego et al. 2014.

Zongle, Lü et al., New incremental clustering framework based on induction as inverted deduction, in: Journal of Systems Engineering and Electronics, 20 (2009) 1132-1143.

Recent Publications

Computer Music. Electronica, Algorithms, Artificial Intelligence

The Gifted Individual and the Right to Education

www.ingramcontent.com/pod-product-compliance
Lightning Source LLC
Chambersburg PA
CBHW051331220526
45468CB00004B/1588